A Wisley Handbook

Propagation from Seed

JIM GARDINER

Cassell

The Royal Horticultural Society

THE ROYAL HORTICULTURAL SOCIETY

Cassell Educational Limited
Wellington House, 125 Strand
London WC2R 0BB
for the Royal Horticultural Society

Copyright © Jim Gardiner 1997

All rights reserved. This book is protected by copyright. No part of it may be reproduced, stored in a retrieval system, or transmitted, in any form or by any means, electronic, mechanical, photocopying, recording or otherwise, without written permission from the Publishers.

First published 1997

British Library Cataloguing in Publication Data
A catalogue record for this book is available from the British Library

ISBN 0–304–32062–5

Line drawings by Mike Shoebridge
Photographs by Neil Campbell-Sharp (pp. 17, 19, 21, 35, 39, 46, 57 (left); Jim Gardiner (pp. 8, 13, 14, 15, 16, 26, 27, 28, 29, 30, 31, 32, 41, 42 (right), 53, 55); Andrew Lawson (p. 51); Harry Smith Horticultural Photographic Collection (pp. 4, 6, 9, 11, 37, 42 (left), 49, 57 (right))
Typeset by RGM, Southport
Printed in Hong Kong by Wing King Tong Co. Ltd

Cover: Pots and seedheads.
Photograph by Jack Townsend/Insight Picture Library
Back cover: Pricking out seedlings.
Photograph by Jim Gardiner
p.1: The flesh of mistletoe berries (*Viscum album*) is sticky, enabling the seed to be distributed by birds.
Photograph by Harry Smith Horticultural Photographic Collection

Contents

Introduction

Seed is by far the most common way by which flowering plants reproduce themselves and, for us, sowing seed is a straightforward and inexpensive method of raising plants for our gardens and containers. Our first introduction to this method of propagation is usually in a garden centre or store, where racks of brightly coloured packets of seed for bedding plants and vegetables are on display. Step-by-step instructions are given on the back of every packet which, if followed, should ensure successful results. Sowing seeds and seeing them germinate is extremely satisfying and often encourages gardeners to try other and more challenging areas of propagation.

Once hooked, gardeners may indulge their hobby further and collect and raise their own seed or take advantage of the seed lists offered by specialist societies. For the really adventurous it is even possible to subscribe to share in the results of a seed-collecting expedition.

All seeds are raised using methods similar to those employed for bedding plants. Your own collected seed or that gathered in the wild may not always give as uniform results as that packaged by the large seed companies, and germination may be erratic. Even with packeted seed things may go wrong and a question frequently asked is: 'Why won't they germinate?' A failure may lead to seed being discarded, but in many cases failure is due to a lack of understanding of the plant and the mechanisms that trigger germination.

This understanding of the way different plants work often holds the key to success. Take for example the story of the dove or handkerchief tree, *Davidia involucrata*, introduced from China in spring 1901. The seeds were embedded in a nut of woody tissue which proved absolutely unbreakable. All sowing methods under the protection of glass were tried and, because of the quantity, the majority of nuts were sown outside so as not to waste them. A year on, nothing sown under glass had germinated but, to everyone's surprise, when those sown outside were inspected, many nuts were showing signs of cracking and within weeks the seeds had

Iris foetidissima is grown for its jewel-bright fruits. Like other fleshy coated seeds they need to be cleaned before sowing

The ripe pods of Himalayan balsam (*Impatiens glandulifera*) explode at the slightest touch, throwing seeds clear of the plant

germinated. It was realised then that the exposure to winter frosts held the key to germination. The seeds sown under glass eventually germinated a year later and after a similar winter treatment.

This book not only explains sowing techniques and aftercare of seedlings, but it will also help you to understand why seeds germinate and the conditions necessary to achieve this. A comprehensive list of garden plants and vegetables commonly raised from seed is included listing specific requirements for particular genera.

Other methods of propagation, from cuttings, layers and divisions, are detailed in a companion Wisley Handbook: *Propagation from Cuttings*.

Understanding Seeds

Understanding how plants grow is an important part of successful gardening. The life cycle of a garden plant starts with a seed (or spore in the case of a fern).

WHAT ARE SEEDS?

The seed is formed from the fertilisation of the female part of the flower by pollen received from the male part. Seeds range enormously in size – from large coconuts and chestnuts to dust-like begonia and orchid seed – and in the quantity produced.

A seed is an extremely efficient structure for reproduction and multiplication. Many seeds retain their viability for years, resistant to desiccation and extremes of temperature. Those that are short-lived are generally produced in great quantities, up to several million in the case of orchids. A protective seed coat encloses an embryo plant and, in most instances, a food store to start the embryo into growth and to sustain the young plant until it is capable of manufacturing its own food. The composition of the food reserve is variable: it may be rich in fats (olive and peanut), in carbohydrate (cereals) or in proteins (peas and beans). The nature of the food reserve has a direct effect on how long seeds can be stored and remain viable.

A seed is classified according to where the food is stored. If the food store remains within the seed coat below the ground the seed is described as endospermic. Endospermic seed results in hypogeal germination – where the food store remains within the seed coat below ground level. Non-endospermic seed has food stored in well-defined seed leaves, or cotyledons, which emerge above ground in what is known as epigeal germination.

METHODS OF DISPERSAL

As plants are rooted literally to the spot, how do seeds move away from the parent plant? Seeds are contained within a fruit, either succulent or dry. Succulent fruits are fleshy and often brightly coloured, making them attractive to animals and birds which eat them. The seeds pass through the digestive tract, their viability unaffected. Mistletoe has a fleshy fruit which is also very sticky.

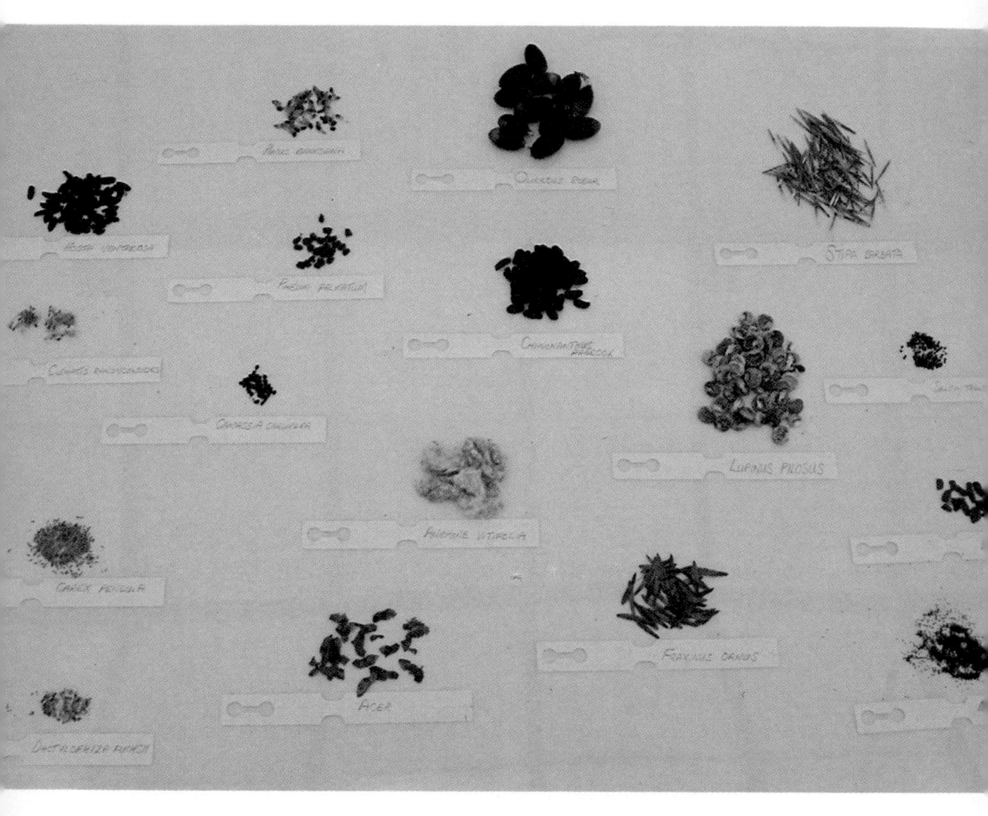

Seeds from different plants vary greatly in terms of appearance, some bearing appendages that aid distribution. Here is a selection gathered from the RHS Garden at Wisley

This stickiness enables it to be carried on the feet and beaks of birds.

Hard-coated seeds are also carried by animals. Squirrels gather nuts, such as acorns and walnuts, often burying them for their winter store. Some seed coats bear spines and hooks which become attached to fur and feathers. Many dry seeds have evolved to be dispersed predominantly by the wind. For instance, maple and ash seeds have developed wings, while those of dandelions and clematis bear plumes.

Other seeds are capable of floating in water: the loose fibrous outer coat of coconuts make them buoyant, and sedges trap air within a membranous envelope. Plants like lupins and impatiens release their seeds explosively, throwing them clear of the parent plant.

Sycamore (*Acer pseudoplatanus*), like other maples, bears winged seeds held in bunches and often referred to as 'keys'

WHY RAISE PLANTS FROM SEED?

For the gardener, many plants can be raised cheaply and successfully from seed, without any particular skills or the need for special facilities or equipment.

Raising new cultivars or seeds collected from the wild widens the gene pool. Any new plants raised are evaluated for improved characteristics, such as better flower quality or resistance to pests and diseases.

Packeted seeds are often described as F1 hybrids, which means they are the result of crossing two true-breeding species. This is an expensive and time-consuming process, for pollination and harvesting must be done by hand annually. The result is uniform, vigorous and often more desirable plants. However, if seed is collected from an F1 cross, known as an F2, it will not produce the same uniformity in the next generation. Other packeted seeds are described as 'open-pollinated' (OP) as the seeds are the result of natural pollination. The results of open pollination lack the uniformity of F1 hybrids. A few plants can produce seed without fertilisation taking place and the resulting seedlings will be identical to the mother plant. This is known as apomixis. Apomictic species are found in several genera including *Sorbus*, *Citrus*, and *Poa*.

A great advantage of propagating from seeds is that they can be moved easily and cheaply around the world. Some countries have plant health restrictions so it is important to get in touch with the relevant agricultural or plant health authority before importing seed. The legal controls relating to the collection of seed from native habitats are discussed on page 12. Seed sold commercially has to conform with minimum standards of viability, cleanliness and purity.

There are comparatively few disadvantages in raising plants from seed rather than by vegetative means, from cuttings for example. The disadvantages are seedling variation; woody plants raised from seed take longer to flower than from cuttings; and the loss of certain characteristics, such as variegation, which cannot be transferred by seed.

SEED BANKS

Some horticultural and botanical establishments have set up seed banks to conserve the world's flora for future generations. One such example is The Royal Botanic Gardens, Kew. Such seed banks can only store dry seed. Some seeds will survive for many years if kept sufficiently cold (−18°C/0°F) and dried to a five per cent moisture content. However, many tropical tree seeds and other large seeds (oaks, palms and many aquatic plants) cannot be stored in this way. For these the possibility of storing seed embryos at low temperatures is being investigated. For practical methods on how you can store seeds, see page 16.

The seeds of *Populus tremula* and other poplars are covered in fluffy white 'cotton'. Sow the seeds straight away without removing the 'fluff'

Sources of Seed

Commercial seed companies offer a wide range of popular plants to the gardener. Catalogues are available either through your local garden centre or can be sent for direct. A selection of seeds from the major seed companies is available from garden centres or through your local horticultural club. Seed companies advertise in the garden press and *The RHS Plant Finder* offers a list of over 70 seed stockists. In addition, there are societies and specialist seed companies which between them offer an extremely wide selection. You will have to join the appropriate society to take advantage of its seed list, but this alone is often worth the annual subscription.

SEED COLLECTING IN THE GARDEN

It is important to gather seed in the right way, even if collecting is confined to your own garden. Always take an indelible pencil, labels and sealable bags – paper bags and envelopes are the cheapest, but avoid worn envelopes as small seeds will find any holes in the corners. Alternatively fold your own envelopes out of newspaper. Do not use polythene bags as seed will probably rot if left in them uncleaned for long.

Most seed is harvested before it is ripe. If a section of stem bearing a partially ripened seedhead is cut, the seeds will eventually ripen. This is described as harvesting 'in the green'. Many members of the buttercup family are collected and sown in the green, among them *Anemone nemorosa*, *Helleborus*, *Hepatica* and *Ranunculus*. Pansies and violas, too, need to be collected in the green; if left to ripen the capsules will split open when touched and release the seeds. Balsams (*Impatiens*) are probably the worst for doing this and will explode in all directions. In this case, the green capsule is cut off complete with the stalk and placed in a paper bag. Always label your seed as you collect it.

COLLECTING SEED FROM THE WILD

This can be rewarding, but should always be carried out in a responsible way. Without care, further decline in numbers and possible extinction of wild plants may result. In Britain a number of species are protected under the Wildlife and Countryside Act, 1981 and its subsequent amendments, and should not be collected. Similar legislation exists in much of Europe, Australia, Canada and the USA. In addition to these legal requirements there is a worldwide agreement known as Convention on International Trade in Endangered Species of Wild Fauna and Flora (CITES) in force in more than 100 countries. It covers the movement of plants across national borders, and is controlled and regulated by a licensing system.

If you propose to collect seed from the wild, then it is important to contact the relevant organisations well beforehand. The Department of the Environment, Endangered Species Branch, Tollgate House, Houlton Street, Bristol BS2 9DJ, will provide you with information. Advice can also be obtained from The Royal Horticultural Society (free to members), The Royal Botanic Gardens (Kew and Edinburgh) or the Alpine Garden Society.

When collecting in the field, note the following data for each seed type: altitude, aspect, associated species and soil type, and catalogue this information under a collection number which should appear on the plant label.

SEED PROVENANCE

The original geographical source of seed is known as seed provenance. This information is important when a species is widespread. Variations in structure, adaptations to soil and climate, resistance to pests and diseases, may characterise plants of a particular locality. For instance, seeds from a tree growing in a warm area or at low altitude will be difficult to establish in a colder region or at a higher altitude, despite the fact that both extremes are within the natural distribution of the species.

Many suppliers give specific provenance details when selling plants, particularly exotic woody species, of questionable hardiness. For instance, eucalyptus seeds collected in sub-alpine areas of Victoria, the mountains of New South Wales and Tasmania, are preferable for growing in Britain than those collected from warmer climatic regions, which would be more difficult to establish.

DRYING AND CLEANING

Take care to keep seeds from different plants separate when cleaning and always keep them labelled. Most seed should be partially dried, preferably on a window-sill in the sun or in an

Once the flowerhead has dried the seed is easier to extract. Always work over a sheet of clean white paper

The cones of *Abies koreana* will disintegrate on the tree if left, so collect them intact and store indoors in an open box to catch the seeds

airing cupboard. Place seeds of one type in a box or seedtray lined with clean brown or white paper. Alternatively, if the seed heads are collected on the stem, bundle the stems together and insert them, head first, into a paper – not polythene – bag. Tie the neck of the bag and hang it upside down in a dry room until the seeds have been released. Once dried break up the seed capsules between finger and thumb over a clean sheet of white paper. Pick out the larger pieces of dried plant and separate the seeds from the chaff using a combination of sieving and winnowing (gently blowing over the seeds and chaff: the lighter chaff will be blown away). Aquilegias, delphiniums and primulas are examples of seeds to be treated in this way.

Larger seeds, such as those of maples (*Acer*) or hornbeams (*Carpinus*), need little cleaning other than rubbing or cutting off the wings. The cones of conifers can be dried in a warm room. The scales of pine cones will open and release the seed, while fir cones (*Abies*) will disintegrate. Cedars, on the other hand, require soaking in warm water for 48 hours before the scales will open.

The seeds of willows (*Salix*) and poplars (*Populus*) with their hairy, cotton-like coverings, and feathery seeds, typified by clematis, do not need cleaning. Tease them out on the surface of the compost, cover them with grit and water in.

It is easy to extract seeds from a quince (*Chaenomeles speciosa*); just cut a ripened fruit in half and pick out the seeds

Many Australian plants grow in areas where bush fires occur. Eucalypts and callistemons are two examples where the seeds are locked inside very hard capsules. In the wild the seeds are only released after the capsules have been heated by fire. This can be simulated by heating the capsules in a frying pan. When you hear one go pop, remove the capsules from the heat. Once they have cooled, small seeds should be released. Alternatively cut the capsules open with a knife.

Many plants have their seeds contained with a succulent fruit. Apples, pears and quinces, for example, can be cut through and the seed extracted easily. *Berberis* and *Physalis* can be messy as the flesh needs to be squashed, or macerated, against a sieve wall and rinsed in running water. To clean off the final remnants of flesh, place the seeds in a jar of warm water and leave for a few days. The viable seeds sink, leaving the non-viable seeds floating on the surface. Larger fleshy seeds of cotoneasters, hawthorns (*Crataegus*) and *Iris foetidissima*, for example, need similar treatment to *Berberis*, but are less messy.

Alternatively fleshy seeds can be roughly cleaned, mixed with three parts their volume of moist sand and placed in a polythene bag in the fridge. The seed and sand mixture can also be placed in a large pot in the garden until you are ready to sow. Cover the pot with fine wire mesh to prevent mice from eating the seed.

Magnolias have fleshy seeds, but with oily seed coats. Remove the oily film by soaking the seeds in warm water to which non-biological detergent has been added.

STORING SEED

Most seeds are stored dry and so avoid fungal diseases and rotting. Once the seeds of trees and shrubs have been thoroughly dried they can be sealed in polythene bags and labelled. For short-term storage, keep seeds in a cool dry place where the temperature remains even. For longer-term storage, place the bags in a fridge (but not a freezer). Seed can be kept for several years at 3–5°C (35°F) and low humidity, however, the viability of oily seeds like those of magnolias is limited to nine to twelve months.

Seeds of flowers and vegetables can usually be kept for about two years. Store them in paper or cellophane bags as they need a higher moisture content than those of woody plants. If the conditions are cool and moist, individual packets of seed can be stored in large, sealed polythene bags.

Seeds that will not germinate if dried, such as acorns, must be stored moist. The majority are large and are mixed with moist vermiculite, sand or peat. Place the mixture in a sealed polythene bag, keep at 5°C (40°F) and sow within two or three months. If the temperature is higher, germination may occur in the bag.

Seed alstroemeria is packed into a cellophane envelope. Label each packet with the plant name and collection date, and store them in a cool place in a sealed polythene bag until sowing time

— Factors Affecting Germination —

Factors affecting germination include the length of time the seed has been in the packet, as viability is reduced over time and dormancy. Certain environmental factors play a vital role in the successful germination of seeds and these are water, temperature, light and air.

Water All seeds take up water. The rate of uptake is determined by the type of seed coat and other outer coverings. Conversely, if there is too much water, germination will be suppressed due to lack of oxygen.

Temperature Germination speed is controlled by temperature. For the majority of temperate plants, germination occurs between 8–18°C (47–65°F), and for tropical plants between 15–24°C (60–75°F).

For successful results sow seeds of hellebores as soon as they are ripe, in the condition known as 'in the green'

Light Most seed germinates equally well in the light or in the dark. Although light triggers germination, the convention is to cover all but the finest with compost or gravel so the seed is in constant contact with water held within the compost. Artificial lighting is sometimes used commercially to control germination. In such cases low-intensity fluorescent daylight tubes, switched on for eight hours a day, are recommended.

Air Seed germinates in air, which if reduced or excluded, will inhibit germination. See the above note on water.

DORMANCY

If all the factors described above are present, but the seed fails to germinate, then it is either non-viable or dormant. Dormancy means the seed has developed mechanisms to prevent it from germinating until favourable conditions prevail. Although dormancy benefits plants in the wild, it creates problems for the gardener. Temperate plants have three types of dormancy: embryo, seed-coat and chemical. In some cases a combination of dormancies is exhibited.

Embryo dormancy

If seeds germinate in autumn seedlings will be exposed to harsh winter weather and may not survive. However, if the same seeds are placed under glass and kept at 15°C (60°F), they would fail to germinate. This is because once the seeds take up moisture, they must undergo a period at low temperatures (2–5°C/34–40°F) before they can germinate. This natural process can be simulated by soaking seeds in water for 48 hours, draining them and storing in polythene bags in a fridge for three weeks to three months. Large quantities of seeds should be mixed with moist grit or vermiculite, placed in a polythene bag and sealed. This process is known as stratification and is one of the commonest methods of breaking dormancy, especially for trees and shrubs (see page 43).

If you do not want to control the time of germination, sow seeds of trees or shrubs in autumn, place in a cold frame over winter and germination will take place in spring.

Lilies and peonies are two examples of plants that need to be sown in warmth, followed by a period of cold before seed will germinate. Other plants, that set seed early in the year, such as daphnes, hellebores and ranunculus, are sown in the green before a germination inhibitor has developed. Germination either takes place quickly and young plants are established enough to survive the winter, or they germinate evenly the following year after being subjected to winter temperatures.

Seed-coat dormancy

This dormancy is found commonly in plants of hot, dry regions. Water does not readily penetrate the seed coat and can only do so once it has been degraded by abrasion (scarification) or by the action of soil organisms. This results in staggered germination, a means of reducing competition between seedlings.

Seeds can be artificially scarified by putting them in a screw-top jar with a little sharp sand or gravel. Screw on the top and shake the jar to abrade the seed coats so water can penetrate. In some cases seeds can be nicked with a knife or the seed coat rubbed with sandpaper. This is time consuming and an easier alternative is to steep one part seeds in three parts warm (not boiling) water and leave for 24 hours. Check the seed has swollen before sowing.

Seed-coat dormancy can also be broken by heat. Forest fires

The seeds of *Callistemon* have an extremely hard coat cracked open by fire in their native Australia. In cultivation, nick the seed coat with a knife to extract the seeds

crack open tough capsules and cause pine-cone scales to open, releasing seeds that may have remained dormant for several years. The seeds drop on the scorched ground ready to germinate with the onset of rain in a competition-free environment.

Chemical dormancy

Chemicals which suppress germination are often present in fleshy seed coats, or within the seed itself. Fleshy fruits are eaten by animals and the seeds pass through their digestive tracts, eventually to be voided. The journey through the animal's body will remove the inhibiting chemicals and the seed will germinate, given suitable environmental conditions.

Daphnes, magnolia and many plants of the rose family, including roses, *Cotoneaster*, *Sorbus* and *Crataegus*, combine chemical dormancy with embryo dormancy. To overcome the chemical dormancy, collect these seeds with fleshy coats before they have ripened (in the green). Clean the seed as described on page 13 and store in polythene bags in the fridge before sowing for two to three months, after which germination occurs once temperature increases.

SEED PRIMING

Some specialist seed firms offer primed seeds. Priming means given a limited amount of water to the seeds to start off the germination process and so ensuring all seeds reach the same stage of development. The primed seed is then dried and packeted. Once sown, primed seed germinates more evenly and quickly, and often at a lower temperature than normal. Chitted or pre-germinated seeds are also available. In this case seeds are germinated to the point where the first root is emerging before they are despatched to the gardener in sealed packets or small plastic containers. Chitting is used where high germination temperatures are required, such as for cucumbers, or where germination is uneven.

Most vegetables and hardy annuals are raised exclusively from seeds. These can be sown in a seedbed or directly in their growing positions

Basic Techniques

The basic sowing techniques and aftercare of seedlings can be adapted to suit most seed types and situations. Variations used in raising particular groups of plants such as alpines, glasshouse plants and herbaceous plants are dealt with on pages 34–49.

SOWING SEED IN OPEN GROUND

Vegetables and many annuals are routinely raised in a seedbed in the garden. This is also the cheapest way of raising trees and shrubs on a large scale, or where no aftercare facilities exist. Obviously you have less control over conditions in the garden than when raising plants under glass. In addition you will need to protect seeds and seedlings from birds and rodents. Weeds germinating at the same time as the seedlings must be identified and rogued out carefully (the desirable seedlings will be found in greater numbers).

Preparing a seedbed
Ideally choose a protected site for your seedbed. If the position is exposed it is worth erecting a windbreak of synthetic netting or

wooden laths around the bed. Decide on the area of your seed bed. A bed 1–1.2 m (3–4 ft) wide and surrounded by a path means that cultivation is easy as you can reach all parts from the side, ideal for raising vegetables and woody plants. Alternatively, if you do not want surrounding paths you can edge the seed bed with a framework of boards 10–20 cm (4–8 in) high and 1–1.2 m (3–4 ft) wide. The advantage of these systems is that you concentrate your efforts on cultivating and improving soil fertility only where the seeds are sown.

To prepare the seedbed, dig over the ground during autumn so the soil can weather and break down over winter, especially important if it is heavy. Even a light sandy soil is worth digging in autumn as there are many other gardening jobs to carry out in spring. If the ground is not in regular cultivation or is heavy, double digging is advisable, otherwise single digging will suffice. However you dig, incorporate well-rotted organic matter, such as rotted farmyard manure or leafmould (and grit in heavy soil) as you work. The addition of leafmould into the seedbed greatly assists the future development of tree and shrub seedlings by providing them with the myrorrhiza (type of fungus) necessary for good root and shoot growth.

In spring, lightly fork over the ground to break up any remaining clods and roughly level off the ground. This also encourages any weed seeds to germinate. These can be hoed or sprayed off before sowing. Apply a dressing of an organic or inorganic fertiliser containing phosphate (P) and potash (K). Inorganic fertilisers tend to be released more quickly than organic ones.

Before sowing, rake over the soil, removing large stones. Small seeds (carrots, onions and hardy annuals) need a fine tilth, while for larger seeds, such as beans, a rougher finish is satisfactory. Tread light soil before raking, but never tread wet or clay soils as the structure will be damaged and clay soils will become compacted.

Before you sow

Once the seedbed is ready, bear in mind the following points before you sow as they will affect germination:

■ The soil must be damp, but not waterlogged.

■ Soil temperature must be at least 5°C (40°F), but between 8–15°C (46–60°F) is better. If you have a heavy soil or wish to raise the soil temperature, spread perforated polythene film or a polythene sheet over the ground. Remove these once you have sown to avoid damaging the emerging seedlings.

■ Always sow at the correct depth: aim for a covering approximately twice the diameter of the seed.
■ Follow the instructions on the seed packet. Timing will vary according to the crop, when you want it to mature and where, geographically, you live. Trees and shrubs that need stratifying are sown in late autumn; most vegetables, annuals, and trees and shrubs that do not require stratifying are sown in spring.

Sowing techniques
Large seeds are space sown singly using a dibber to make holes 5 cm (2 in) deep. If the hole depths vary, uneven germination will result. Cover the seeds and firm the soil with the back of the rake to ensure good contact between seed and soil.

Sowing in drills Small seeds are sown in drills, using a garden line to mark the position of the drill. The depth of the drill depends on the size of the seed. Seeds are best sown on a still, dry day. Sow

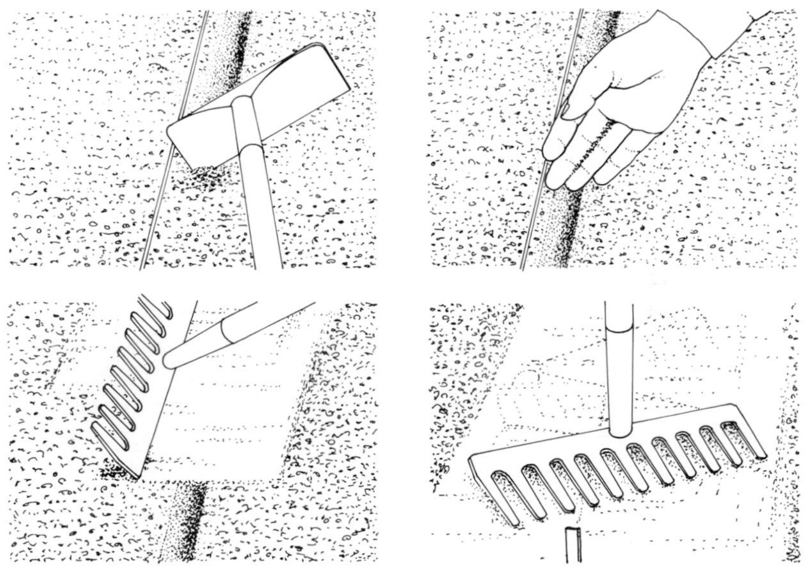

1 Draw out a shallow V-shaped drill using the corner of a draw hoe, with the blade held against a garden line. The drill can also be made using the tip of a cane or pressing a wooden board into the soil
2 Sow seeds thinly and evenly along the drill
3 Cover the seeds with soil using the back of the rake
4 Tamp the soil using the back of the rake to firm the surface. Label the rows with the name of the crop and the sowing date. Another method of covering the seeds is to shuffle along the drill. This not only pushes the soil back into the drill but firms it as you go

as thinly as possible for this not only reduces the need for thinning, but also the risk of fungal attacks. (However, old seed should be sown more thickly as the percentage of viable seed will be lower than with fresh seed.) Either sow direct from the packet, gently tapping it so the seeds fall out from along the crease, or take a pinch of seeds between the thumb and forefinger and sprinkle them along the drill. Aim for a density of two or three seeds per 2.5 cm (1 in) run, up to 10 cm (4 in) apart for plants whose final spacing is 25 cm (10 in). Mix fine seeds with silver sand to facilitate thin, even distribution.

Some seeds are available in pelleted form. Pelleted seed is expensive but easy to space sow. The protective coat quickly takes up moisture and disintegrates, so pellets need to be sown soon after the airtight pack is opened or they will become soft and mushy. Sow two pellets per station.

If the soil has a high clay content, make a shallower drill and cover the seed with proprietary seed compost. Alternatively fill the drill with soil, then cover the ground with horticultural fleece or newspaper until after the seeds have germinated. This reduces the chance of soil capping and makes it easier for the seedlings to emerge. Seeds of trees and shrubs that take several months to germinate or are fine, such as birch (*Betula*), benefit from a covering of fine (3 mm/$\frac{1}{8}$ in) grit to a depth of 12 mm ($\frac{1}{2}$ in).

Broadcasting Prepare the seedbed well in advance and ensure it is weed-free. Scatter the seed over the soil and lightly rake it in. Rake the bed again at right angles to the first raking and label the crop. Water in the seed using a can with a fine rose or, on a larger scale, a sprinkler with a fine mist-like spray. If the ground is heavy, water it before sowing, avoiding the areas where you are likely to stand. Broadcasting is used for hardy ornamental annuals, carrots, mustard and cress, green manures, wild-flower mixtures and grass seed.

Protecting seeds and seedlings

If rabbits are a problem, then surround the seedbed with fine wire mesh netting, buried to a depth of 15 cm (6 in). Set traps in old clay land drainage pipes to catch mice. Birds can be deterred by stringing humming tape or a network of taut black cotton 3–5 cm (1–2 in) above the soil.

To catch problems early, inspect the crop regularly. Look out for seedling diseases and disorders and any sign of pests, and take appropriate measures.

Frost can kill or damage seedling crops, particularly vegetables, broad-leaved trees and ornamental annuals. On a small scale cover the seedbed with newspapers when frost is forecast or, more effectively, with horticultural fleece. The mechanically minded can set up a spray irrigation system triggered to cut in when the temperature falls to 0.5 °C (33 °F). Such a system protects seedlings down to −6 °C (21 °F).

Ensure emerging seedlings do not dry out, particularly on light, sandy soils during windy, sunny days. Water as necessary and apply liquid fertiliser regularly at the recommended rate.

Thinning

Thin seedlings before they become too spindly. Make sure the soil is moist before thinning, then either nip out surplus seedlings at soil level or, preferably, pull them out. Firm in the remaining seedlings and place the thinnings on the compost heap. The space between seedings varies with the crop. For plants requiring 20 cm (8 in) or more, it is better to make one, two or even three thinnings rather than leaving tiny seedlings isolated and vulnerable.

SOWING SEED UNDER COVER

The facilities for sowing seed in a protected environment vary considerably, from a small, unheated propagator on the windowsill or cold frame in the garden, to an environmentally controlled propagator in the glasshouse. Even a pot standing in a saucer and covered with glass will provide some protection. Shade seeds with newspaper if they are in a sunny situation to reduce temperature fluctuations. Obviously the facilities you need depend on the plants you want to raise and in what quantities. For instance, a small cold frame is sufficient for raising a limited number of trees, shrubs or hardy annuals, while house plants and bedding plants require a propagator.

Cold frames can be wooden or aluminimum with a hinged or sliding light (glass lid). Remember these small structures can easily overheat during sunny weather unless well ventilated and shaded. Conversely, they lose heat quickly, so when frost is forecast cover the frame with hessian, coconut matting or carpet, but remove if the temperature rises above zero during the day.

Propagators can be heated or unheated. If unheated, the larger the canopy the better the protection against sudden changes in temperature. Thermostatistically controlled propagators give greater control over temperature, and this can be varied as

A cold frame is an invaluable aid to raising plants from seed, offering protection to seedlings and safe standing room for pots of seeds overcoming dormancy

necessary. Most provide a temperature range between 7–30°C (45–85°F) with the heat provided either by a mat or cables. Always ensure the electrical supply is properly fused and plugged into a circuit breaker, thus protecting both you and your equipment from any short circuit or overload.

Containers

Small quantities of seed can be raised in pots. Plastic pots are widely available and come in different colours. Black absorbs more heat than paler colours – useful when raising seed early in the year. Pots also come in different shapes. Square pots fit closer together and take up less space than round ones. Clay pots are still widely used for specific plant groups, notably alpines. Long toms (deep clay pots) are ideal for individually sown tree seeds and plants which develop a long tap root. Clay pots are porous so dry out quicker than plastic. For this reason clay pots are often plunged in moist sand. Plastic pots are not porous, so need careful watering to avoid overwetting the compost. The cheapest pots are thin-walled and moulded into a block of small, individual units or cells which last for one or two seasons.

Seedtrays or half-trays are used for sowing large quantities of seed, but a more even distribution of seedlings and easier handling

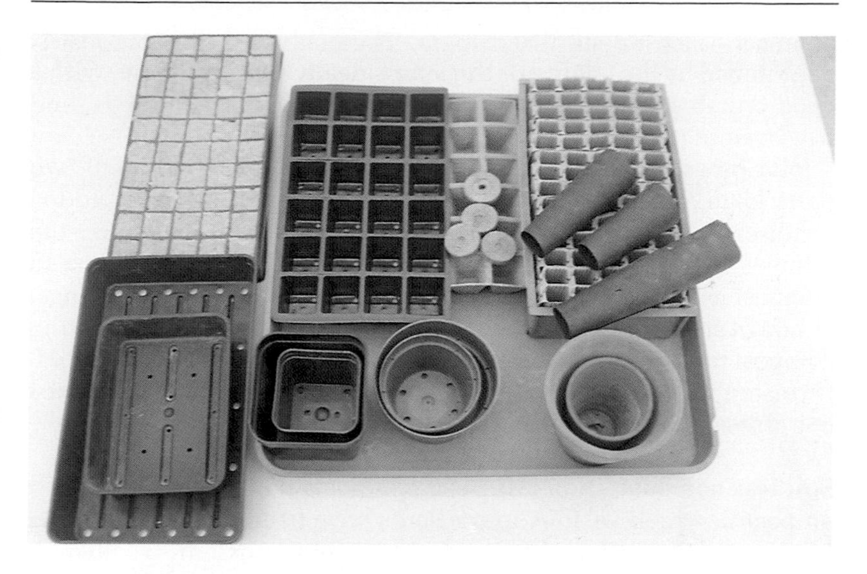

Trays, pots and cellular modules are all suitable for raising seeds, depending on the quantities involved. The long pots are particularly useful for starting off sweet peas and other plants that produce tap roots

can be achieved by sowing seed in several small pots. For this reason moulded plastic or polystyrene cellular trays, so-called 'plug trays', are popular. These units can hold between 60 and 500 seedlings. They are pre-formed and generally stand on a base. Use them either for direct sowing or for growing on seedlings.

Composts

This term is widely used in the UK to describe the growing medium in which seeds are raised or plants are grown in containers. It is also applied to the product of the compost heap. In this book the term refers only to the growing medium, for which there are a multitude of recipes.

A proprietary compost is best for raising the majority of seeds. This can, if necessary, be modified to suit particular plants. Composts are basically classified as being either soil-based or soil-less, although the ingredients may vary. Whatever the type of compost it should be used, preferably, within a month of purchase as the higher the temperature and moisture, the greater the quantity of nutrients available.

Soil-based compost In the 1920s work at the John Innes Horticultural Research Institute produced the first uniform compost based on sterilised loam. The quality of the compost is determined by the quality of the loam: ideally slightly acidic, with a good crumb structure, free of pests, disease and weed seeds, and free-draining but without drying out too fast.

John Innes (JI) seed compost, as it is known, is made up of two parts loam, one part peat, one part sand (by volume) plus added fertiliser. JI potting composts are numbered 1 to 3, the higher the number denoting the greater the amount of fertiliser. JI seed compost is used not only for sowing but for pricking out seedlings, although strong-growing seedlings are better moved into JI potting compost no. 1 in late spring or early summer when growth is rapid. If you are not using the compost immediately store it in a cool place out of the sun.

Soil-less compost Soil-less seed composts have long been based on peat to which various ingredients have been added, including sand, perlite, vermiculite and, to a lesser extent, rockwool. Leafmould or sphagnum moss have been used instead of peat for many years, especially for raising rhododendrons. More recently, composted bark, coir, straw, animal waste products and municipal refuse have been used as peat alternatives. Fertilisers are added in a soluble or slow-release form, the latter often resin or plastic coated to control the rate of release into the compost.

From left to right, top: peat, silver sand, 3 mm ($\frac{1}{8}$ in) grit; *centre*: sphagnum moss, vermiculite; *bottom*: John Innes seed compost, soil-less seed compost

The advantages of seed raising in soil-less compost are: it is lightweight; clean to use; sterile; and difficult to over-firm. The disadvantages are: it is difficult to re-wet once allowed to dry out; mosses and liverworts may form on the surface if seed is slow to germinate. A corollary to this is that young plants have difficulty in establishing in heavier soils if grown entirely in soil-less composts compared to those grown in soil-based composts.

Sowing techniques

Before you start to sow, get all your materials and equipment together. Make sure you have an adequate supply of compost and that it is sufficiently moist. Test this by taking a clenched handful of compost, open your hand and it should remain moulded together; if water oozes out, the compost is too wet. Containers must be clean and dry. Fill the container as described and place the pot or seed-tray on a clean sheet of paper to catch any seed. You are now ready to sow.

Ready for use on this well-equipped potting bench are compost, trays, pots, presser boards, grit, vermiculite, sieve, labels, pencil and, of course, seeds

Left: Always over-fill the container, firm any corners and tap the pot or tray once or twice on a solid surface to ensure there are no air gaps
Right: Add more compost so it is mounded above the top of the container, and strike off excess compost using a sawing action

Left: Gently firm the compost with a pressing board, slightly smaller than the container
Right: The compost level should be about 12 mm ($\frac{1}{2}$ in) below the rim

Always sow seed thinly. Sow about half the seed as described, then turn the container through 90 degrees and sow the remainder. Cover the seed to approximately twice its depth with compost, compost plus fine (3 mm/$\frac{1}{8}$ in) grit or vermiculite, or fine grit or vermiculite alone. Label each pot with the name and sowing date. Where germination can take two or more seasons, record the sowing information in a book giving name of crop, date and description of the technique. Leave columns for germination date and notes on aftercare.

Water the pot or tray either from below, by standing in a water bath so the moisture is taken up by capillary action, or if grit covers the seed, use a watering can with a fine rose. However, never start or stop watering over the top of the container, or drips will fall on the prepared surface. Do not overwater or soak too long, as waterlogged compost will encourage damping-off diseases if germination is rapid.

Left: Sow seed directly from the packet
Right: Alternatively it can be sown from the slightly cupped palm of the hand, gently tapping the wrist so that seed runs evenly onto the compost

Sowing fine seed

Above: Before sowing, sieve a layer of compost over the prepared surface. This gives fine seed good contact with the compost and slows drying out. Label before sowing to avoid disturbing the compost

Above, right: You will find it is easier to sow thinly if you first mix the seed with silver sand. Sow the mixture evenly over the pot or tray in two directions as described above

Right: Do not cover fine seed, and water by standing the container in a water bath

Aftercare

Once seeds are sown and watered, place containers in a cold frame or propagator. Inspect at regular intervals for signs of germination: weekly in a cold frame; daily in a heated propagator. For seeds that need overwintering, it is a good idea to plunge containers up to their rims in sand to reduce changes in moisture content between waterings. Germination temperatures range from 8–24°C (47–75°F), with hardy plants germinating at the lower end, vegetables midway, and bedding and glasshouse plants at the upper end.

Once germination has taken place and the seedlings are large enough to handle, prick them out into trays or pots. Use a compost similar in character and moisture-holding ability to the sowing mixture. Fill the containers as described above, page 29.

Pricking out

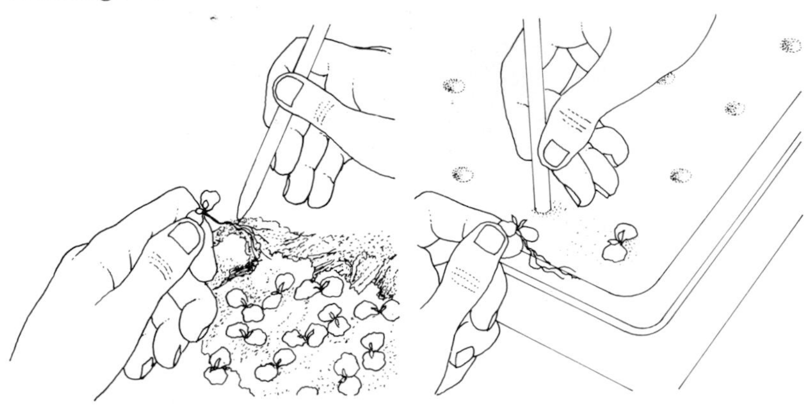

1 Ensure the container of seedlings is moist before carefully upturning the container and gently tapping out the contents onto the work surface. Tease out the seedling root systems with a dibber or plant label
2 Remember always to handle the seedlings by a seed leaf, never by the stem. Use a dibber to make a hole large enough to take the entire root system

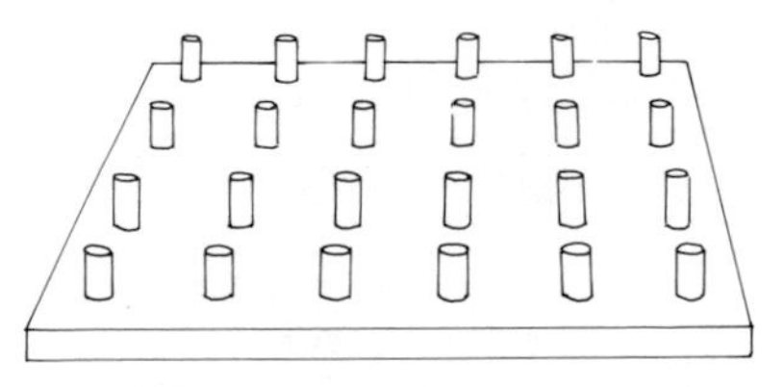

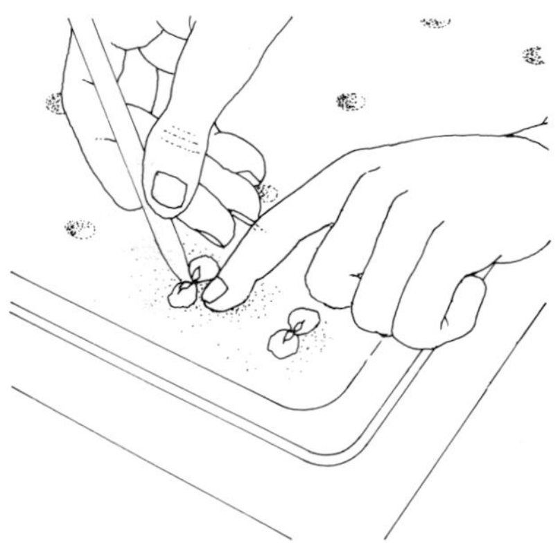

Alternatively make this simple gadget to fit your seedtrays: attach between 24 to 48 pegs to a presser board, and press into the compost to make planting holes
3 Ease the roots into the hole and gently firm the compost round the roots with the aid of the dibber

An alternative method is to loosen the compost by tapping the pot or tray against the bench and teasing out the seedlings with a dibber. However, the method illustrated causes less damage to the tiny root systems. Once the seedlings have been pricked out, label with the name and date of pricking out, water in and stand in a propagator or cold frame. Place any surplus seedlings back in the original container and add more compost.

During the first few days after pricking out ensure the seedlings are shaded from direct sunlight and apply a copper-based fungicide against damping-off disease.

Prick seedlings out as soon as they are large enough to handle

Hardening off

This means weaning the seedlings off the environmentally controlled conditions within the propagator or cold frame so they will be able to survive in their final, unprotected location. A few days after pricking out, increase air movement by raising the frame lid or increasing ventilation within the propagator. Remove any shading unless sunlight is intense or shade-loving plants are being raised. The hardening-off process also involves spacing out seedlings, either by moving the pots further apart or potting on seedlings from trays to individual pots. The ambient temperature must also be reduced. This can be achieved not only by increasing ventilation but also by moving plants from the propagator to the glasshouse bench, cold frame or under a cloche. If seedlings are already in a cold frame, remove the lid. However, shut the frame tight and cover it with hessian if frost is forecast, particularly where bedding plants are concerned. To help avoid fungal attacks do not over-feed young plants and keep the compost on the dry side. If in doubt apply a preventative spray of copper-based fungicide.

Specific Techniques

Certain groups of plants require modifications to the basic sowing techniques described in the previous chapter and these are detailed below.

HARDY ANNUALS

Annuals prefer a sunny situation and fertile, moisture retentive soil. In general, sow hardy annuals directly in open ground during spring. Autumn sowing can achieve good results, particularly of *Limnanthes*, *Centaurea* and *Viscaria*, although success does depend on a mild winter. As the soil can be dry in late summer it is sufficient to cultivate just the top 5 cm (2 in) rather than break up the soil in the conventional way. Rake the soil to a fine tilth and sow. For a spring sowing, prepare the ground in late autumn and sow in drills as described on page 23. If you are making a display of several different hardy annuals, arrange the drills of each plant type at varying angles. Follow the spacing guidelines given on the packet. Broadcast sowing can result in a glorious mixture if plants are selected carefully.

The thickness of sowing depends on the type of seed and, to a lesser degree, on the soil conditions. Seed sold by commercial companies has a high percentage viability, while that of your own

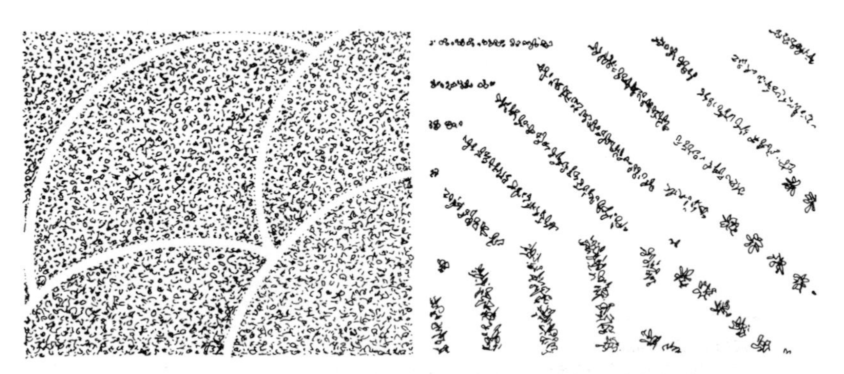

Left: Sow annuals in bays drawn out with a sharp stick or marked with a trickle of silver sand
Right: Within each bay sow seeds thinly in drills

Half-hardy annuals and bedding plants can be raised from seed sown in the protection and warmth of a propagator

collected seed may well be lower. It will take two or three thinnings to arrive at the recommended final spacings. By thinning over a period, seedlings give each other mutual protection and, if necessary, can be moved to fill gaps. Use a narrow-bladed (bulb) trowel to transplant seedlings and water them in immediately. On thin, hungry soils, liquid feed seedlings with a balanced fertiliser.

Biennials are raised from seed in the same way as hardy annuals, except that they are sown in nursery rows 30–40 cm (12–16 in) between rows and thinned to 15–23 cm (6–9 in) apart. Plant them out in their final positions as recommended on the packet.

HALF-HARDY ANNUALS

Bedding plants are raised from seed indoors as described on pages 25–26. Use plastic pots or seedtrays filled with a proprietary soil-less seed compost. For best results seed should be covered with vermiculite. (Unlike compost vermiculite will not compact, so seedlings can be pricked out more easily, and a better air to moisture ratio is maintained.) Fine seeds such as those of begonia, tobacco plant (*Nicotiana*) and petunias are mixed with silver sand and sown as described on page 31. Begonia seeds are now available pelleted for easier handling. Sowing dates range from mid-winter to early spring for summer bedding, and from late spring to midsummer for winter bedding. Temperatures vary between 15–25°C (60–77°F) and are recommended on the packet.

Following germination, seedlings should remain in the propagator until pricked out. If they are growing strongly, increase ventilation to slow down growth. Prick out seedlings singly into trays, modular units or individual pots. However, lobelia is usually pricked out in small clumps. Apply a copper fungicide as recommended to prevent damping off.

ALPINES

Always try to sow alpine seeds fresh or in the green, although this is not always possible with seed obtained commercially. Store seed that cannot be sown immediately in the fridge. A lower percentage of old seed will germinate and germination will be slower than with fresh seed, possibly up to three years. Alpines are generally sown in autumn or early winter, and preferably placed in a cold frame to overwinter. However, very fine seed, such as that of *Ramonda* and *Haberlea* should be sown in spring in a propagator kept at 12–15°C (55–60°F).

Clay pots are recommended for high alpines which naturally inhabit free-draining, rocky or stony mountainous regions, while plastic pots or half-pots are suitable for alpines growing at lower altitudes where moisture does not run off immediately.

Because good drainage is vital, I recommend a JI seed compost with added sharp sand, free from all traces of clay particles (this may entail washing the sand before use). As JI composts contain some lime, they are not suitable for raising alpines needing acidic conditions, such as Asiatic gentians. In such instances a soil-less seed compost should be used. To fill a clay pot, first cover the drainage hole with a piece of broken pot (concave surface uppermost). Then add 2.5 cm (1 in) grit, mounded in the centre before filling with compost and firming the surface, as described on page 29.

Seeds can be space sown or sown thinly from the packet. Cover the seed with sharp sand, label and water in. Cyclamen is an

Before sowing *Gentiana acaulis* and other European gentians cover the compost with a layer of fine grit. Sow the seeds on the grit and water in to bring them in contact with the underlying compost

example of seed that can be space sown easily. Soak old cyclamen seed for three or four hours in tepid water to which a few drops of non-biological detergent have been added. This will help break down the waxy seed coat and allow water to be taken up more readily.

Many alpine seeds are too coarse to be surface sown, yet too fine to be covered with grit. In these instances (primulas and European gentians, for example) cover the compost with a fine covering of 3 mm ($\frac{1}{8}$ in) grit before sowing. Then label and water; the water will take the seeds down to the compost. Grit can be extracted from sharp sand by sieving to collect the desired particle size. Chicken grits, Cornish grit or Chichester grit (for lime-loving plants) are alternatives available from aquarium stockists.

After sowing place the pots in a cold frame. Do not plunge high alpines as drainage will be impaired. However, alpines of lower altitudes and those sown in the green can be plunged in moist sharp sand free of clay particles, (see page 31) to within 2.5 cm (1 in) of the rim.

If the seed fails to germinate the following spring, leave the pots in the cold frame for a further season or two. If moss becomes a problem, water with a dichlorophen solution at the recommended rate.

After germination, prick out the seedlings when they are large enough to handle. If germination is sporadic, leave them over winter and prick out the following spring when more seedlings may have appeared. Do not prick out after August, as the shortening days means that seedlings will not be well enough established to overwinter successfully. Prick out seedlings into JI seed compost, but use a potting compost for vigorous alpines or a soil-less compost for acid lovers. Place a layer of 3 mm ($\frac{1}{8}$ in) grit round the neck (lower stem) of seedlings to provide extra drainage. Cushion plants can have thin plates of rock carefully positioned around their stem. Alternatively they can be planted into tufa, a porous limestone. You will need a masonry drill to make a 5–8 cm (2–3 in) deep hole in a block of tufa. Fill the hole with a mixture of tufa fragments and compost, carefully insert the seedling with the aid of a dibber and water in.

Once pricked out, return the seedlings to the cold frame, or glasshouse bench. At this stage high alpines benefit from being plunged in sharp sand to reduce temperature and moisture variations in the compost. During their establishment, alpine seedlings need plenty of ventilation to keep the air moving (a fan can assists with this), and good light but not direct sun.

HERBACEOUS PLANTS

Although many herbaceous plants grown in gardens are cultivars which need to be propagated vegetatively to come true to type, others, *Geranium*, *Dictamnus*, *Lysimachia* and *Incarvillea* for example, include many species and so can be successfully raised from seed. Variation can occur within a batch of seedlings, often to the raiser's advantage as illustrated by the many beautiful seed raised forms of *Helleborus orientalis*. To this end, tag any plants that show good colour so seed can be collected later.

Raise herbaceous plants as described on pages 29–33, using a JI or soil-less seed compost in pots or half-trays. Fine seeds need no covering, see page 31, and herbaceous gentians can be sown as European gentians, see on page 37.

The seeds of many herbaceous plants ripen during the second half of the year, so clean and store before sowing in late autumn. Unless the seeds are fine, cover them with fine grit. After sowing stand containers in a cold frame. Germination should take place in spring. If you do not have a cold frame, store seeds dry in the fridge.

Many herbaceous perennials can be raised from seed, including hardy geraniums, hellebores and darmera

Sow them in spring, cover with vermiculite and place in a propagator at 15–21°C (60–70°F) where they should germinate within four weeks.

Ideally sow spring-flowering herbaceous plants immediately they ripen, or just before, and raise them in a cold frame. Rates of germination are fairly good. However, it is advisable to over-winter pots in a cold frame where, invariably, a secondary germination occurs. You can either prick out the seedlings as they germinate and return the pot to the cold frame or wait until following spring.

Double dormancy

Herbaceous peonies and trilliums exhibit what is known as 'double dormancy' where the root systems germinate a year before the shoots develop. As soon as the seeds ripen, usually in late summer, remove them from their capsules and sow in JI seed compost. The large seeds of peonies are pressed into the compost, covered with more compost and then topped with fine grit. Treat trilliums the same way, though some gardeners prefer to cover seeds with 6 mm (¼ in) bark chippings. Label and water, then place the pots in a cold frame, plunged in sand. After the period of winter cold, the root system develops during the first spring and summer. The following year, after the second cold period, the shoots appear. Germination time can be reduced by a year by mixing seeds with three times their volumes of moist peat or vermiculite in late summer. Place the mixture in a polythene bag, seal and store in a fridge for two months. Then remove to a heated room or glasshouse kept at a temperature of 17–21 °C (65–70 °F) for two months, by which time root should have formed. With great care, tease out the individual seeds and pot up in JI potting compost no. 1. Peonies are potted individually and covered with grit, while trilliums are potted up in small groups and covered with bark. Shoots should appear the following spring.

Lilies The majority of lily species are sown in the conventional way, see page 29. However some species (those with hypogeal germination, belonging to sections Archelirion and Martagon) exhibit double dormancy and should be treated as described above for peonies and trilliums. If you are unsure as to which group your lily belongs, sow in autumn onto JI seed compost, cover with fine grit, label, water and place in a cold frame. If germination has not occurred within six months, wait another year.

AQUATIC PLANTS

Although the majority of aquatic plants are raised vegetatively, a few are propagated by seed. Collect seed of submerged aquatics, like *Hottonia* and *Nymphea odorata*, and floating plants, such as *Aponogeton* and *Ludwigia*, as soon as it is ripe. Sow immediately in plastic pots filled with JI seed compost, with added charcoal to prevent the waterlogged compost becoming too acidic. Cover the seed with fine grit and submerge in a water bath, so the pot is covered by about 5 cm (2 in) water. Keep the water at 15 °C (60 °F).

Sow marginal aquatics, such as *Caltha*, *Lysichiton*, *Sagittaria* in a similar way, but with water reaching only half-way up the pot. Cover the pot with a piece of glass to keep the atmosphere humid. The air temperature should be between 13–16 °C (55–60 °F). After germination prick out small clumps into aquatic pots. Gradually harden off the seedlings and place them outside in a pond during summer.

BULBOUS PLANTS (excluding lilies)

Bulbous and cormous species (not cultivars) in cultivation can easily be raised from seed. The limiting factors are the time taken from germination to flowering – from one to six years – and the number of viable seeds. The method is similar to that used for raising alpines, see page 36.

Seeds of South African bulbous species, such as *Dierama pulcherrimum*, should be sown in spring

For dwarf bulbs (those under 15 cm/6 in) use half-pots or pans; standard pots are better for large bulbs whose root systems need the extra depth. Sow fresh seed in autumn or early winter in JI seed compost and cover seeds with fine grit. Water in and place pots in a cold frame.

South African bulbs *Babiana*, *Dierama*, *Gladiolus* and *Moraea*, and some Californian bulbs such as *Calochortus* do not need a cold period to trigger germination. These can be sown in spring and raised in a glass house or propagator. Store the seed in the fridge until you are ready to sow.

Germination usually occurs in spring or early summer; if not wait a further 12 months. Prick out seedlings of vigorous growers into individual pots either singly or in small clumps. Seedlings of less vigorous plants can be kept in the same pot until the following spring, or even the spring after that. During the intervening period, in the early part of the growing season, water regularly and apply a low nitrogen liquid feed every two or three weeks to build up the bulb size. Stop feeding once the foliage starts to die back.

TREES AND SHRUBS

As fresh seed is important for reliable germination, you will be confined mostly to collecting seed from your garden or locality. Seeds from a cultivar or from a plant growing near closely related species are likely to produce variable seedlings. The variations

Left: Acorns of *Quercus coccifera* and other oaks deteriorate rapidly and must be sown in autumn as soon as they are ripe
Right: The scarlet fruits of deciduous *Euonymus europaeus* burst open to reveal the brilliant orange seed covering known as an aril. Remove the aril before sowing

may or may not be an improvement on the parent. Seedlings will only come true to type if they are raised from apomictic species (see page 10) or from plants growing in comparative isolation from related species.

The majority of woody plants produce flowers and ripe seed in the same year, so seeds can be sown in autumn. Large seeds, such as acorns (*Quercus*) and sweet chestnuts (*Castanea*) will not germinate if they dry out, so sow them immediately they have ripened in autumn.

Certain fleshy seeds, such as *Daphne* and *Sorbus*, if collected in the green, give quick and even germination within a few months of fertilisation (see pages 12 and 18). Others, for example *Arbutus* and *Skimmia*, need to remain on the plant until they ripen before sowing. The fleshy fruits of magnolias and euonymus burst open in autumn to display the brightly coloured, ripened seeds.

Mock oranges (*Philadelphus*), deutzias, forsythias and lilacs (*Syringa*) produce dry fruits or capsules. These, like the pods of *Colutea*, *Laburnum*, and other leguminous trees and shrubs, change colour from green to yellow, brown or black when ripe.

Collect the capsules just before they split along the edge and release the seed.

Rhododendrons and the majority of the heather family (*Ericaceae*) produce masses of very fine, dust-like seeds in capsules, which ripen from late summer into autumn. Leave them on the plants for as long as possible before you pick them. Then place the capsules on a sheet of paper or in a paper bag in a warm, dry place until they are about to split and the seed can be collected.

Many conifers flower in spring and produce cones the same year. These include firs (*Abies*), *Chamaecyparis* and Douglas fir (*Tsuga*), while pines (*Pinus*) and often cypresses (*Cupressus*) take two or more years to produce ripe cones.

For details of collecting and storing, see pages 12–16; dormancy factors affecting germination are covered on pages 17–20; sowing in open ground is described on pages 21–25.

Sowing techniques

The majority of trees and shrubs can be raised in a cold frame if a cold period is necessary to break dormancy. Alternatively, seed (other than large seeds like acorns) can be stored in the fridge and sown in spring. Sow seeds in pots or trays; long toms, sweet-pea tubes and root trainers are useful for raising deep-rooting plants or where minimal root disturbance is important. Sow in JI seed compost if germination is likely to take more than nine months, but for fine seed a soil-less seed compost is more suitable as germination is generally rapid. Use a soil-less ericaceous compost for all woody ericaceous plants including rhododendrons.

Immediately after harvesting sow large seeds in individual containers or space sow in a deep seedtray. Push seeds into the compost, and cover with compost to twice their depth with a layer of fine grit, label, water and place in a cold frame. If seedlings emerge during winter, section off part of the cold frame and protect seedlings from frost with hessian, coconut matting or old carpet. Prick out space-sown seedlings into individual containers filled with JI potting compost no. 1.

Stratification The majority of hardy trees and shrubs need to undergo a cold period before they germinate. The simplest way to achieve this is to sow cleaned seed in autumn as soon as it is ripe in a JI or soil-less seed compost. Cover the seed with fine grit, label and water. Plunge the pot into sand in a cold frame. Check at regular intervals during winter, especially after a mild period, for seedling emergence. If this has occured, place containers in a section of the frame where they can be covered with hessian or old carpet when frost threatens. If greater control over germination is required, store the cleaned, dry seed in a polythene bag, seal and keep in a fridge. In the middle of winter remove the bagged seed and add sufficient water to cover the seed (you may prefer to use a jar). Return to the fridge for 48 hours, then drain off excess water. If you are dealing with small quantities, return the moist seed to the polythene bag, reseal and replace in the fridge. Soaking primes the seeds, so that once the cold period has been satisfied – from three weeks to three or four months – germination will take place. Mix large quantities of seed with moist grit or vermiculite. This will separate the seeds and ensure there is adequate air space around them during storage. Check weekly for signs of germination. See the table on page 56 for more specific plant information.

At the first signs of germination, remove the seeds carefully to a pot or pan filled with compost and distribute them evenly over the compost surface. Cover lightly with compost topped with grit or vermiculite. Label, water and place in a propagator or on a glasshouse bench at a temperature of 10–15 °C (50–60 °F). Prick out the seedlings into individual pots filled with JI seed compost or potting compost no. 1, or a soil-less equivalent.

If the seed has failed to germinate after four months from priming, it is either dormant or not viable. To find out which, sow, cover with fine grit and place in a cold frame. If the seed is viable, germination should occur the following spring (that is, 18 months after priming).

A number of shrubs, including hydrangeas, deutzias, callistemons and lavenders, do not need a cold period. In these instances, store cleaned seeds in polythene bags until it is time to sow in spring in a JI or soil-less seed compost. Cover lightly with vermiculite, label, water and place in a cold frame or glasshouse or, for a more rapid result, in a propagator at 10–15 °C (50–60 °F). Prick out seedlings into pots or seed trays. Nine months later you can move them into individual pots for planting out the following spring.

Rhododendrons and other ericaceous shrubs such as *Enkianthus*, *Kalmia*, and *Gaultheria* all have fine seeds. Collect the ripe capsules at the end of the year, clean and store in paper packets which in turn are placed inside polythene bags and sealed. Store in a fridge until spring. *Gaultheria* fruits are soft and fleshy, so clean them as described on page 15 before storing. In spring fill half-pots or half-trays with soil-less ericaceous seed compost. Sieve the final 12 mm ($\frac{1}{2}$ in) layer of compost and firm. Before sowing sieve a thin layer of compost over the surface but do not firm. Sow thinly, do not cover the seed, label and water from below (see page 30). Place the container in a propagator or mist unit at a constant temperature between 10–15°C (50–60°F). If you are using a propagator, mist inside at regular intervals. After germination allow the seedlings to grow on and prick them out the following spring.

GLASSHOUSE AND CONSERVATORY PLANTS

Most glasshouse plants are grown in one of three temperature regimes:

- Cool: 5–10°C (40–50°F).
- Warm: 10–15°C (50–60°F).
- Tropical: 18°C (65°F) and above.

You will need to know in which temperature band the parent plant grows as this gives a guide to the germination temperatures required. However, fine seeds, of cinerarias and calceolarias for example, need high temperatures (21°C/70°F) for germination, but once established will grow in cool conditions. Try to sow fresh seed, particularly for tropical plants, as the older the seed the longer the germination takes and the lower the percentage that germinate.

Glasshouse plants are raised from seed indoors as described on pages 25–33. Achimenes, cinerarias, gloxinias, primulas and streptocarpus have fine, dust-like seeds which should be sown in half-pots, pans or seedtrays filled with a soil-less seed compost, as described on page 31, and placed in a propagator. Check daily for signs of germination, moisture loss and disease. Spray with a copper fungicide against damping off. Prick out seedlings into individual pots or modular units, and harden off.

Larger seeds, such as *Browallia*, Swiss cheese plant (*Monstera*) and passion flowers (*Passiflora*), are sown in JI seed compost or a soil-less equivalent and covered with either sharp sand or vermiculite. Citrus plants and coffee have large seeds which need to be cleaned and, ideally, sown fresh. Packeted seed should be sown immediately you receive it or germination rates can be disappointing. Space sow these seeds on JI seed compost and cover with fine grit. Place in a propagator at 18–21°C (65–70°F). Fresh seed should germinate within five weeks. Prick out the seedlings into individual pots and keep in cool to warm temperatures. *Acacia* species are also raised in this way but, for even germination, first soak the seeds as described on page 18.

Seeds of *Passiflora racemosa* require temperatures of between 18–25 °C (65–80 °F) to germinate reliably

Palm seed Once ripe, this has a very short period of viability so should be sown as fresh as possible. Remove the fleshy outer covering as it contains germination inhibitors. Partially cleaned seed can be soaked in warm water and kept in a warm place for a few days while the fermentation process lifts off any fleshy remains. Date palm seeds can be abraded to assist germination, see page 19.

Sow in JI or soil-less seed compost in seedtrays, clay pots or root-trainers. The latter are ideal as they are deep and allow the tap root to develop unchecked. Lay large fruits, such as coconuts, on their sides and bury them to approximately half their depth; depress smaller seeds into the compost and cover them to their depth with fine grit. Water in. The ideal germination temperature for most palms is 24–30°C (75–85°F), but never below 21°C (70°F). Germination times vary considerably, from four weeks to two years, but most fresh seeds germinate in two or three months. Wean the seedlings off the high temperatures as most palms prefer either tropical or warm regimes. Move those palms not already in individual containers into pots. Palms need free drainage so are better raised in clay pots rather than plastic.

CACTI AND SUCCULENTS

Although the seed of many species can remain viable for several years, for best results sow fresh seed. Sow as early as possible in spring, so by winter the seedlings will have grown large enough to survive through winter. Use JI seed compost with extra grit in a 10 cm (4 in) half-pot or pan. With very fine seed, like that of *Mammillaria*, cover the firmed compost with sharp sand and sow onto this. Large seeds, such as *Astrophytum*, should be first soaked in warm water for three or four hours to speed up germination before sowing on the firmed compost and covering with fine grit.

Aim for a night temperature of 21 °C (70 °F), increasing during the day by natural radiation from the sun to 32 °C (90 °F). Germination will take place at lower (warm) temperatures but will be slower. Maintain high humidity by covering the pot with a piece of glass or using a propagator, but remove condensation from the glass or propagator lid daily, so large droplets do not land on the compost. Unless the seed is covered with grit always water from below just as the compost starts to dry out, and keep a watchful eye for damping off. Mammillarias should germinate between ten days and eight weeks.

Seedlings can be pricked out from three months after germination, though some species should be left longer. Try to prick out the same year as sowing, but if this cannot be done before August it is better to leave the job until the following spring. In the meantime ensure the seedlings are in good light, and wean them off the high humidity levels provided during germination. Prick out seedlings into a mixture of JI potting compost no. 1 and grit (an additional one part by volume or more can be added) for improved drainage. Knock the seedlings out of the pot so the maximum amount of root is transferred. To further facilitate pricking out the tiny seedlings use tweezers or a 'fork' made from an old plastic label if the finger and thumb method proves too clumsy. Space the seedlings about 3–5 cm (1–2 in) apart and top-dress with fine grit. Water in, shade from strong sunlight at first and ventilate freely.

VEGETABLES

Most vegetables are raised from seed sown directly in open ground or under protection in containers. The seeds are sold as F1 hybrids, often vacuum-packed, or as open pollinated (OP). Seeds are also available pelleted, in tape strips, chitted (sprouted) and as pre-germinated seed in gel which can be fluid sown. All these formats reduce the need to transplant. Seed kept longer than two or three seasons is unlikely to germinate.

Sowing in open ground

Prepare the ground as described on pages 21–22. The optimum soil temperature varies with the crop, but between 8–13°C (47–55°F) should give good results for most crops.

Sowing in drills This is the most common method for raising vegetables, see page 23. Either space the seeds evenly along the drill or sow them at stations. For the latter method, place two or three seeds at double the final spacing in the row. This is common practice for peas, beans and many root crops including carrots, parsnips and beetroots. For peas make a flat-bottomed drill 10–23 cm (4–9 in) wide.

To save space sow a fast-maturing crop, like lettuce or radish, in between a slow one such as parsnips; the faster maturing crop will act as a marker indicating the position of the slow-germinating seeds. Another method of raising vegetables that need a long growing season, brassicas in particular, is to sow them in seed beds and plant them out later. Space sow seeds 5 cm (2 in) apart, in rows 20 cm (8 in) apart. Make sure the soil is moist before transplanting.

Broadcasting This is suitable for sowing green manure crops, mustard and cress, an early carrot crop or for 'cut-and-come-again' salad crops. After thoroughly preparing the ground and ensuring it is weed-free, sow the area as evenly as possible. The sowing rate is generally high, with salad crops sown at 12 g per m² (½ oz per sq yd). Rake the seed in two directions, the second at right angles to the first. 'Cut-and-come-again' crops are cut with a knife or scissors when plants reach 15 cm (6 in) in height. They will resprout and can be cropped several times in this way.

Large seeds These include marrows, beans and sweetcorn. Sow two seeds in a hole made with a dibber. Cover the seeds and firm gently to ensure they are in contact with the soil. Cover marrow and pumpkin seed with a jam jar or cut-off plastic bottle to increase the soil temperature. After germination thin out the weaker seedlings.

Under protection

Vegetables can be raised in pots, seedtrays, modular units or soil blocks in a protected environment, ranging from a window-sill in the house to a cloche or cold frame, propagator or glasshouse. Sow

Runner bean seeds are large enough to be space sown in holes made with a dibber

as described on page 25, in a JI or soil-less seed compost. Cover the seed with a thin layer of vermiculite or sieved compost, label and water. Optimum temperature is between 13–15°C (55–60°F), but it should not fall below 8°C (47°F). Prick out into seedtrays as soon as the seedlings are large enough to handle. Space seedlings 2.5–5 cm (1–2 in) apart. Alternatively prick out seedlings into individual pots or modules before hardening off and planting out.

Another method is to sow direct into modules. This is particularly useful when only a few plants are being raised. It also has the advantage of reducing the number of moves a seedling has to undergo before planting out. Plants raised in modules get a good start and suffer minimal root disturbance. There is a wide range of modules available including Plantapak, polystyrene cellular trays, biodegradable modules and soil blocks. In most cases sow one seed per module. For leeks and onion sow four or five seeds per cell. Pelleted seeds are easy to sow singly. A simple way of sowing ordinary seeds individually is to lift one off a board with a damp watercolour brush; the seed sticks to the brush and falls off on contact with the compost. After sowing, cover seeds thinly with compost, label and water in. Gradually harden off the seedlings before planting out in their final location.

Garden Plants from Seed

Name	Month to sow/ preparation	Sowing method	Germination Temp.	Germination Time (weeks)
ANNUALS, BIENNIALS, BEDDING PLANTS				
African marigold F_1 types (HHA)	1–2 indoors	pots/trays/modules; lightly cover with compost	20–25°C (68–77°F)	1–2
Ageratum F_1 types (HHA)	3 indoors	pots/trays; surface sow; lightly cover with vermiculite	18–25°C (64–77°F)	2
Alyssum (HHA)	3 indoors	pots/trays; cover with vermiculite	13–15°C (55–60°F)	2
Amaranthus (HHA)	3 indoors	pots/trays; surface sow; lightly cover with vermiculite	18–21°C (64–70°F)	2
Antirrhinum F_1 types (HHA)	2 indoors	pots/trays; surface sow	18–21°C (64–70°F)	2–3
Aster (HHA)	3 indoors	pots/trays; cover with vermiculite	18–21°C (64–70°F)	2–3
Begonia semperflorens (HHA)	1–3 indoors	pots/trays; surface sow; cover with fleece to maintain humidity. Sow primed/pelleted seed in modules	20–25°C (68–77°F)	2–4
Bellis (biennial)	6–7 cold frame	pots/trays; cover with vermiculite	15°C (60°F)	3
Brassica (HA) (ornamental cabbage, kale)	6–7 indoors or outdoors	pots/modules/seed-bed drills; 20–25 mm ($\frac{3}{4}$–1 in) deep; transplant from seed bed	15°C (55°F)	1
Calendula (HA)	3–4 outdoors	drills; 6–12 mm ($\frac{1}{4}$–$\frac{1}{2}$ in) deep	10–15°C (50–60°F)	1–2
Clarkia (HA)	3–4 outdoors	drills; $\frac{1}{2}$ in (12 mm) deep	8–10°C (46–50°F)	3
Cosmos (HHA)	2–3 indoors	pots/trays; cover with vermiculite	15–18°C (60–64°F)	1
Dahlia (HHP)	2–3 indoors	pots/trays; cover with vermiculite	15–20°C (60–68°F)	2–3
Dianthus barbatus Sweet William (biennial)	4–5 outdoors (early sowing) indoors	seed bed drills for transplanting; sow 12 mm ($\frac{1}{2}$ in) deep pots/trays	12–15°C (54–60°F)	2–3
Eschscholzia (HA) (California poppy)	3–4 outdoors	drills; 12 mm ($\frac{1}{2}$ in) deep	10–12°C (50–54°F)	2

Name	Month to sow/ preparation	Sowing method	Germination Temp.	Germination Time (weeks)
Gazania (treat as HHA)	9–10 indoors	pots/trays; overwinter for May flowering	15–18°C (60–64°F)	1–2
Impatiens F_1 type (treat as HHA) (busy lizzie)	2–3	trays; surface sow; cover with 3 mm (⅛ in) vermiculite, then with fleece to maintain humidity	21–24°C (70–75°F)	2–3
Kochia (HHA) (burning bush)	3 indoors	pots/trays; ½ in (12 mm) deep	20°C (68°F)	2
Limonium (HHA) (statice)	2–3 indoors	pots/trays; lightly cover with compost	15–20°C (60–70°F)	2–3
Lobelia (HHA)	1–3 indoors	pots/trays; lightly cover with vermiculite	15–21°C (60–70°F)	2–3
Mesembryanthemum (HHA) (Livingstone daisy)	2–3 indoors	pots/trays; lightly cover with compost	15–20°C (60–68°F)	2–3
Nicotiana (HHA) (tobacco flower)	3 indoors	pots/trays; 6–12 mm (¼–½ in) deep	15–20°C (60–68°F)	2–3
Pelargonium F_1 types (HHP)	1–3 indoors	trays, or modules if using treated seed	21–24°C (70–75°F)	2–3
Petunia (HHA)	2–3 indoors	pots/trays; surface sow; do not cover	18–22°C (64–74°F)	2–3

As *Begonia semperflorens* produces very fine seed, look out for primed or pelleted seeds which are easier to handle and give good germination results

Name	Month to sow/ preparation	Sowing method	Germination Temp.	Time (weeks)
Polyanthus (P) (*Primula veris*)	5–7 indoors	pots/trays; surface sow; do not cover; eratic germination at higher temp.	15–18°C (60–64°F)	2–3
Salvia splendens (HHA)	2–3 indoors	pots/trays; lightly cover with vermiculite	18–21°C (64–70°F)	2–3
Tagetes (HHA)	3–4 indoors	pots/trays/modules; lightly cover with compost	18–21°C (64–70°F)	1–2
Verbena (HHA)	3 indoors	pots/trays; surface sow; cover with vermiculite	20–25°C (68–77°F)	3–4
Viola tricolor (P) (pansy)	5–7 autumn flowering 9–10: spring flowering	pots/trays/modules; 6–12 mm (¼–½ in) deep	15–22°C (60–74°F)	2–3
Wallflower (biennial)	6 outdoors	seed bed drills; space sow 12 mm (½ in) deep	15°C (60°F)	2
(*Erysimum cheiri*)	7–8 cold frame	pots/trays/modules; 6–12 mm (¼–½ in) deep; cover with compost	15°C (60°F)	2

ALPINES

Name	Month to sow/ preparation	Sowing method	Germination Temp.	Time (weeks)
Androsace	9–11 stratification cold frame	pots; sow on grit; water in	8–10°C (46–50°F)	10
Asperula	2–4 cold frame	pots; sow on grit; water in	10–12°C (50–54°F)	3–4
Aubrieta	6–8 cold frame	pots/modules; lightly cover with vermiculite	10–15°C (50–60°F)	2–3
Campanula	2–4 cold frame	pots; surface sow; lightly cover with sand	10–15°C (50–60°F)	3–6
Codonopsis	2–4 cold frame	pots; surface sow; very fine covering of sand	10–15°C (50–60°F)	3–4
Corydalis	2–4 cold frame	pots; sow on grit; water in	10–15°C (50–60°F)	2–4
Dianthus	2–4 cold frame	pots; surface sow; lightly cover with compost or grit	10–15°C (50–60°F)	3–4
Dryas	9–11 stratification cold frame	pots; cover with grit	8–10°C (46–50°F)	10
Draba	9–11 stratification cold frame	pots; cover with grit	8–10°C (46–50°F)	10
Gentiana (gentian)	8–11 stratification cold frame	pots; sow fresh seed on grit; water in	8–10°C (46–50°F)	10–52
Hepatica	7–9 stratification cold frame	pots; sow fresh seed on grit; water in	8–10°C (46–50°F)	10–52
Incarvillea	2–4 cold frame	pots; surface sow; lightly cover with vermiculite	10–15°C (50–60°F)	4
Lewisia	7–9 stratification cold frame	pots; sow fresh seed on grit; water in	8–10°C (46–50°F)	12–52

Name	Month to sow/ preparation	Sowing method	Germination Temp.	Time (weeks)
Linum	2–4 cold frame	pots; surface sow; lightly cover with compost	10–15°C (50–60°F)	4–5
Oenothera	2–4 cold frame	pots; cover with grit	10–15°C (50–60°F)	4–6
Omphalodes	2–4 cold frame	pots; cover with grit	10–15°C (50–60°F)	4–5
Oxytropis	2–4 cold frame	soak for 24 hrs in warm water prior to sowing; pots; cover with grit	10–15°C (50–60°F)	4–6
Phyteuma	2–4 cold frame	pots; sow on grit; water in	10–15°C (50–60°F)	4
Pulsatilla	9–11 stratification cold frame	pots; cover with grit	8–10°C (45–50°F)	12–52
Ramonda	2–4 indoors	pots; do not cover; place under mist in a closed case	10–15°C (50–60°F)	3–4
Saxifraga	9–11 stratification cold frame	pots; sow on grit; water in	8–10°C (45–50°F)	10

Pulsatilla vulgaris seeds need to undergo a cold period, known as stratification, before they will germinate

Name	Month to sow/ preparation	Sowing method	Germination Temp.	Germination Time (weeks)

HERBACEOUS PERENNIALS

Name	Month to sow/ preparation	Sowing method	Germination Temp.	Germination Time (weeks)
Achillea (yarrow)	1–4 indoors	pots; surface sow; lightly cover with vermiculite	15–20°C (60–68°F)	1–2
Aconitum (monkshood)	1–3 fridge	pots; surface sow; cover with vermiculite	15–18°C (60–64°F)	4–6
Alstroemeria	sow as soon as ripe, double dormancy (see page 18)	mix seed with vermiculite in a polythene bag; keep at 17–21°C (65–70°F) for 2 months; move to a fridge at 3–5°C (37–40°F) for 2 months; replace in warmth, 17–21°C (65–70°F), for 2 months; pot seed, cover with grit, place in a cold frame at 3–5°C (37–40°F) for 2 months		25
Aquilegia	5–6 indoors cold frame	pots; surface sow: cover with vermiculite	15–18°C (60–64°F)	3–4
Astrantia	9–11 stratification cold frame	pots; cover with grit	10–15°C (50–60°F)	12
Coreopsis	1–2 indoors	pots; surface sow; cover with vermiculite	15–20°C (60–68°F)	2–4
Doronicum	5–6 indoors	pots; surface sow; cover with vermiculite	15–20°C (60–68°F)	3–4
Echinops (globe thistle)	1–3 indoors	pots; surface sow; cover with vermiculite	20°C (68°F)	2
Erigeron	1–3 indoors	pots; surface sow; do not cover	15–20°C (60–68°F)	2–3
Eryngium	9–11 stratification cold frame	pots; surface sow; cover with grit	10–15°C (50–60°F)	12
Euphorbia	9–11 stratification cold frame	pots; surface sow; cover with grit	10–15°C (50–60°F)	12
Geranium (cranesbills)	2–4 indoors cold frame	pots; lightly cover with compost; top with vermiculite	15–20°C (60–68°F)	3–4
Helleborus	5–6 (on ripening) stratification cold frame	pots; lightly cover with grit	5–10°C (41–50°F)	40
Liatris	2–3 indoors	pots; surface sow; do not cover	20°C (68°F)	3–4
Lysimachia	2–3 indoors	pots; surface sow; do not cover until germinated, then lightly with vermiculite	15–20°C (60–68°F)	4
Meconopsis (Himalayan poppy)	9–11 stratification cold frame	pots; surface sow; do not cover. Sow *M. grandis* on grit	10°C (50°F)	12
Papaver (poppy)	7–8 indoors cold frame	pots; surface sow; lightly cover with vermiculite	15–20°C (60–68°F)	2
Thalictrum	1–3 indoors cold frame	pots; surface sow; cover with vermiculite	15–20°C (60–68°F)	3–5

Name	Month to sow/ preparation	Sowing method	Germination Temp.	Time (weeks)
Trollius (globe flower)	4–6 indoors cold frame	pots; surface sow; lightly cover with grit	10°C (50°F)	40
Veronica	1–3 indoors cold frame	pots; surface sow; lightly cover with vermiculite	15–20°C (60–68°F)	2–4

BULBS

Name	Month to sow/ preparation	Sowing method	Germination Temp.	Time (weeks)
Allium	1–3 indoors cold frame	pots; surface sow; lightly cover with vermiculite	10–15°C (50–60°F)	2–4
Colchicum	1–3 cold frame	pots; surface sow; cover with grit	10–15°C (50–60°F)	4–8
Crocus	8–10 stratification cold frame	pots; surface sow; cover with grit	5–10°C (41–50°F)	8–12
Dierama (wand flower)	2–3 indoors	pots; surface sow; cover with grit	10–15°C (50–60°F)	6–8
Fritillaria (fritillaries)	9–11 stratification cold frame	pots; surface sow; cover with grit	5–10°C (41–50°F)	10–14
Freesia	4–5: soak in water for 24 hrs	pots; surface sow; cover with vermiculite	18–21°C (64–70°F)	3–4
Gladiolus	2–3 indoors	pots; surface sow; cover with grit	10–15°C (50–60°F)	6–8

Most bulbous and cormous species can be raised from seed. This is *Crocus tommasinianus*

Name	Month to sow/ preparation	Sowing method	Germination Temp.	Time (weeks)
Lilium (epigeal lily)	9–11 cold frame	pots; surface sow; cover with grit	5–10°C (50–60°F)	10–14
Lilium (hypogeal lily)	9 double dormancy	mix seed with vermiculite in a polythene bag; keep at 17–21°C (65–70°F) for 2 months; move to a fridge at 3–5°C (37–40°F) for 2 months; replace in warmth, 17–21°C (65–70°F)for 2 months; pot seed, cover with grit or vermiculite, place in a cold frame at 3–5°C (37–40°F) for 2 months		25
Narcissus	9–11 stratification cold frame	pots; surface sow; cover with grit	5–10°C (41–50°F)	10–14
Iris	9–11 cold frame	pots; surface sow; cover with grit	5–15°C (40–50°F)	10–14

TREES AND SHRUBS

Name	Month to sow/ preparation	Sowing method	Germination Temp.	Time (weeks)
Abies (firs)	1–2 soak for 48 hrs stratification indoors	soak and stratify seed in fridge for 28 days; pots; cover with grit to a depth of 12 mm (½ in)	15–20°C (60–68°F)	2–3
Acacia (mimosa)	1–2 soak for 24 hrs in warm water cold frame	pots; lightly cover with compost topped with grit to a depth of 6 mm (¼ in)	18–24°C (65–75°F)	4–8
Acer (maples)	10–11 cold frame	pots/trays; lightly cover fresh seed with compost topped with grit to a depth of 12 mm (½ in)	5–10°C (40–50°F)	12
Aesculus (horse chestnut)	10 cold frame	pots/trays; cover with 2.5 cm (1 in) compost	5–10°C (40–50°F)	16
Amelanchier (snowy mespilus)	9 clean seed stratification cold frame	pots; cover with grit	10–15°C (50–60°F)	16
Berberis (barberry)	10–12 clean seed stratification cold frame	pots/trays; cover with grit to a depth of 12 mm (½ in)	10–15°C (50–60°F)	2
Betula (birch)	10–11 stratification cold frame	pots/trays; lightly cover seed with grit	5–10°C (40–50°F)	12
Buddleja (butterfly bush)	2–3 indoors	pots; surface sow; lightly cover with vermiculite	10–15°C (50–60°F)	2–4
Callistemon	2–3 indoors	pots; surface sow; do not cover	15–20°C (60–68°F)	2–4
Carpinus (hornbeam)	10–11 stratification cold frame	pots/trays; surface sow; cover with grit to a depth of 12 mm (½ in)	5–10°C (40–50°F)	12

Name	Month to sow/ preparation	Sowing method	Germination Temp.	Germination Time (weeks)
Carya (hickory)	10–11 stratification cold frame	individual pots/Root-trainers; cover with 2.5 cm (1 in) compost	15–20°C (60–68°F)	14
Castanea (sweet chestnut)	10–11 stratification cold frame	pots/trays; cover with 2.5–5 cm (1–2 in) compost	5–10°C (40–50°F)	12–14
Chamaecyparis	1–2 soak for 48 hrs stratification	soak and stratify seed in fridge for 14 days; pots; lightly cover with compost topped with grit to a depth of 6 mm ($\frac{1}{4}$ in)	15–20°C (60–68°F)	2–4
Chimonanthus (winter sweet)	10–11 cold frame	pots; lightly cover with compost topped with grit to a depth of 6 mm ($\frac{1}{4}$ in)	10–15°C (50–60°F)	12
Cistus	2–3 indoors	pots; surface sow; lightly cover with compost or vermiculite	10–15°C (50–60°F)	2–4
Clematis	10–11 stratification cold frame	pots; surface sow; cover with grit to a depth of 12 mm ($\frac{1}{2}$ in)	5–10°C (40–50°F)	12
Cotoneaster	10–12 clean seed stratification cold frame	pots/trays; lightly cover with compost; cover with grit to a depth of 6–12 mm ($\frac{1}{4}$–$\frac{1}{2}$ in)	10–15°C (50–60°F)	12
Cytisus	2–3 soak for 24 hrs in warm water	pots; light cover with compost topped with grit to a depth of 6 mm ($\frac{1}{4}$ in)	10–15°C (50–60°F)	4–8

Left: Berberis seeds have a fleshy coating which needs to be removed before sowing
Right: Sweet chestnuts (*Castanea sativa*) should be sown as soon as they are ripe. The spiny husk protects seeds while they develop and, in the wild, clings to animal fur to aid distribution

Name	Month to sow/ preparation	Sowing method	Germination Temp.	Germination Time (weeks)
Daphne	variable fresh seed stratification cold frame	pots; sow as soon as ripe; lightly cover with compost topped with grit to a depth of 3 mm (¼ in)	10–15 °C (50–60 °F)	variable up to 65
Deutzia	2–3 indoors	pots; surface sow; do not cover	10–15 °C (50–60 °F)	4
Enkianthus	2–3 indoors	pots; surface sow; do not cover; place under mist	10–15 °C (50–60 °F)	4
Eucalyptus	2–3	individual pots/Root-trainers; lightly cover with compost topped with grit to a depth of 6 mm (¼ in)	15–20 °C (60–68 °F)	4
Euonymous (deciduous)	10–11 clean seed stratification cold frame	pots; lightly cover with compost topped with grit to a depth of 6–12 mm (¼–½ in)	5–10 °C (40–50 °F)	12
Fagus (beech)	10–11 stratification cold frame	pots/trays; cover with compost topped with grit to a depth of 12 mm (½ in)	5–10 °C (40–50 °F)	12
Fraxinus (ash)	10–11 stratification cold frame	pots; cover lightly with compost topped with grit to a depth of 6 mm (¼ in)	5–10 °C (40–50 °F)	12
Hippophae (sea buckthorn)	10–11 clean seed stratification cold frame	pots; lightly cover with compost topped with grit to a depth of 6 mm (¼ in)	10–15 °C (50–60 °F)	12
Hydrangea	2–3 indoors	pots; surface sow; do not cover; place under mist	10–15 °C (50–60 °F)	6–8
Ilex (holly)	10–12 clean seed stratification cold frame	pots; lightly cover with compost topped with grit to a depth of 12 mm (½ in)	5–10 °C (40–50 °F)	14
Juniperus (junipers)	10–12 clean seed stratification cold frame	pots; lightly cover with compost topped with grit to a depth of 6 mm (¼ in) Germination delayed above 15 °C (60 °F)	5–10 °C (40–50 °F)	10–12
Juglans (walnuts)	10–12 stratification cold frame	individual pots/Root-trainers; cover with 2.5 cm (1 in) compost	15–20 °C (60–68 °F)	14
Leycesteria	11–12 clean seed stratification cold frame	pots; lightly cover with compost topped with grit to a depth of 6 mm (¼ in)	10–15 °C (50–60 °F)	10–12
Magnolia	9–11 clean seed stratification cold frame	pots/trays; cover with 6 mm (¼ in) compost topped with a similar quantity of grit	10–15 °C (50–60 °F)	10–12
Mahonia	9–11 clean seed stratification cold frame	pots; cover with compost topped with grit to a depth of 6 mm (¼ in)	10–15 °C (50–60 °F)	12–14

Name	Month to sow/ preparation	Sowing method	Germination Temp.	Germination Time (weeks)
Paeonia (tree peonies)	double dormancy	see *Lilium* (hypogeal)		
Pinus (pines)	10–12 *stratification *cold frame	pots; lightly cover with compost topped with grit	10–15 °C (50–60 °F)	12
Quercus (oaks)	10–12 *stratification *cold frame	pots/trays; cover with 12–25 mm (½–1 in) compost	5–15 °C (40–60 °F)	12
Rhododendron	2–3 indoors	pots; surface sow; do not cover; place under mist	15–20 °C (60–68 °F)	4
Rosa species (roses)	9–11 stratification cold frame	pots; lightly cover with compost topped with grit to a depth of 6 mm (¼ in)	10–15 °C (50–60 °F)	12
Sorbus (mountain ash, whitebeam)	9–11 clean seed stratification cold frame	pots; lightly cover with compost topped with grit to a depth of 3 mm (⅛ in)	10–15 °C (50–60 °F)	4–12
Stewartia	1–3 indoors	pots; lightly cover with compost topped with grit to a depth of 3 mm (⅛ in)	15–20 °C (60–68 °F)	4–8
Taxus (yew)	9–11 clean seed stratification cold frame indoors	mix seed with vermiculite in a polythene bag; keep at 20°C (68 °F) for 3 months; move to a fridge at 5°C (40°F) for 3 months; sow, cover lightly with compost topped with grit to a depth of 6–12 mm (¼–½ in) at 10–15 °C (50–60 °F)		30
Ulex (gorse)	2–3 soak for 24 hrs in warm water	pot; lightly cover with compost topped with grit to a depth of 6 mm (¼ in)	10–15 °C (50–60 °F)	4–8
Vaccinium	2–4 indoors	pots; surface sow; do not cover; place under mist	15–20 °C (60–68 °F)	4–6
Viburnum	double dormancy	see *Lilium* (hypogeal)		

GLASSHOUSE AND CONSERVATORY PLANTS

Name	Month to sow/ preparation	Sowing method	Germination Temp.	Germination Time (weeks)
Achimenes (hot water plant)	1–3	pots/trays; surface sow; do not cover but maintain humidity	18–25 °C (65–77 °F)	3
Agave	1–3	pots; surface sow; lightly cover with grit	21–24 °C (70–75 °F)	4–6
Aloe	1–3	pots; surface sow; lightly cover with grit	21–24 °C (70–75 °F)	2–4
Asparagus densiflorus 'Meyeri'	2–5	soak seed for 48 hrs prior to sowing; cover with 18 mm (¾ in) vermiculite	20 °C (68 °F)	4–6
Astrophytum	1–3	pots; surface sow; lightly cover with grit	21–24 °C (70–75 °F)	4–6

*some species do not require a cold period

Name	Month to sow/ preparation	Sowing method	Germination Temp.	Germination Time (weeks)
Begonia	1–12	pots/trays; surface sow; cover with fleece to maintain humidity	20–22°C (68–72°F)	2–4
Browallia speciosa	3 summer 7 winter	pots/trays; lightly cover with compost	18–20°C (64–70°F)	2–3
Calceolaria (indoor types)	8–10	pots/trays; surface sow; do not cover	15–20°C (60–68°F)	2–3
Campanula isophylla	7–1	pots/trays; surface sow	15–18°C (60–64°F)	4–6
Chamaedorea elegans (parlour palm)	collect and clean seed; sow fresh	pots; cover with 6 mm (¼ in) compost topped with grit	24–28°C (75–82°F)	4–5
Chamaerops humilis (Mediterranean fan palm)	collect and clean seed; sow fresh	pots; cover with 6 mm (¼ in) compost topped with grit	24–28°C (75–82°F)	7–8
Cineraria	4–7	pots/trays; surface sow	18–22°C (64–72°F)	2
Coffea arabica (coffee plant)	2–6 soak for 48 hrs	pots/modules; cover with 18 mm (¾ in) grit	20–22°C (68–72°F)	6–8
Coleus (*Solenostemon*)	1–5	pots/trays; surface sow; lightly cover with vermiculite	20–22°C (68–72°F)	2–4
Cyclamen F_1 hybrids	2–4	pots/trays; space sow; cover with compost topped with grit; keep in dark for 3 weeks	15–18°C (60–64°F)	3–4
Exacum affine	2–3	pots/trays; cover with vermiculite	15–20°C (60–68°F)	2–3
Gerbera jamesonii	1–3	pots/trays; cover with compost	18–21°C (64–70°F)	2–3
Howea forsteriana (kentia palm)	collect and clean seed; sow fresh	pots; cover with 18 mm (¾ in) compost topped with grit	24–28°C (75–82°F)	30
Jacaranda mimosifolia	2–3 soak for 48 hrs	pots/trays; cover with grit	20–22°C (68–72°F)	4–6
Kalanchoë	2–3	pots; surface sow; lightly cover with vermiculite	21–24°C (70–75°F)	1–2
Mammillaria	1–3	pots; surface sow; do not cover	21–24°C (70–75°F)	3–5
Phoenix dactylifera (date palm)	collect and clean seed; sow fresh	pots; cover with 18 mm (¾ in) compost topped with grit	24–28°C (75–82°F)	4–5
Primula acaulis	4–6	pots/trays; surface sow; do not cover until germinated, then lightly with vermiculite; eratic germination at higher temp.	15–18°C (60–64°F)	2–3
Ranunculus F_1 hybrid 'Accolade'	8–10	pots/trays/modules; lightly cover with vermiculite	12–15°C (54–60°F)	3–4

Name	Month to sow/ preparation	Sowing method	Germination Temp.	Germination Time (weeks)
Saintpaulia (African violet)	1–12	pots/trays; surface sow; do not cover; maintain high humidity	20°C (68°F)	3–4
Schefflera elegantissima	2–9	pots/trays; cover with vermiculite	20–25°C (68–77°F)	3–4
Schizanthus	10	pots/trays; surface sow; cover lightly with vermiculite	10–15°C (50–60°F)	2–3
Sinningia F_1 hybrids (Gloxinia)	1–2	pots/trays; do not cover; maintain high humidity	18–22°C (64–72°F)	2–3
Solanum capsicastrum (winter cherry)	2–3	pots/trays; prime seed; cover with 6 mm (¼ in) compost	20°C (68°F)	3–4
Streptocarpus (Cape primrose)	1–4	pots/trays; surface sow; do not cover seed; maintain high humidity	20–25°C (68–77°F)	3–4
Torenia fournieri	2–4	pots/trays; cover seed lightly with vermiculite	20°C (68°F)	2

VEGETABLES

Name	Month to sow/ preparation	Sowing method	Germination Temp.	Germination Time (weeks)
Beans, broad	2–5 successional; initially under cloches	drills/dibber; station sow; 3.7–5 cm (1½–2 in) deep	5°C + (40°F +)	2 +
Beans, runner	5–6	station sow *in situ*; 2 seeds per cane 3.7–5 cm (1½–2 in) deep	10–12°C (50–53°F)	2 +
Beetroot	5–6	drills; station sow 18 mm (¾ in) deep	8°C (46°F)	1½
Broccoli, sprouting	4–5	seedbed; drills 18–25 mm (¾–1 in) deep; transplant	5°C + (40°F +)	2
Brussels sprouts	4–5	seedbed; drills 18–25 mm (¾–1 in) deep; transplant	5°C + (40°F +)	2
Cabbage, spring	7–8	seedbed; drills 18–25 mm (¾–1 in) deep; transplant	5°C + (40°F +)	2
Cauliflower	3 cold frame or under cloches	seedbed; drills 18–25 mm (¾–1 in) deep; transplant	5°C + (40°F +)	2
Carrots	3–5	drills; 18–25 mm (½–¾ in) deep; mix with sand or fluid sow	8°C (46°F)	3 +
Celery, self-blanching	3–4 indoors	modules; surface sow, use treated seed	10–15°C (50–60°F)	3
Courgettes	5 indoors	modules; 25 mm (1 in) deep	13–18°C (56–64°F)	1
Leeks	3–5 indoors or under cloches	early sowings in modules, or seed bed under cloches; 2.5 cm (1 in) deep	7°C (44°F)	1½

Name	Month to sow/ preparation	Sowing method	Germination Temp.	Germination Time (weeks)
Lettuce, maincrop	3–6 successional	modules/drills/fluid sow; 12–18 mm (¼–¾ in) deep	10–24°C (50–75°F)	1
Onions, bulb	2–4	modules/fluid sow in drills 12–18 mm (½–¾ in) deep	10–15°C (50–60°F)	1½
Onions, spring	2–7 successional	drills; 12–18 mm (½–¾ in) deep	10–15°C (50–60°F)	1½
Parsnips	4–6	drills; fluid or station sow 2.5–5 cm (1–2 in) deep	10°C (50°F)	3
Peas, maincrop	4–7	flat-bottom or V-shape drill, or bed system; 2.5–5 cm (1.2 in) deep	10°C (50°F)	2
Radishes	3–8 successional	drills; 12 mm (½ in) deep	8°C+ (46°F+)	1½
Sweet corn	5–6 indoors under cloches	pots/modules/station sow; 2.5–4 cm (1–1½ in) deep	10°C+ (50°F+)	2
Sweet pepper	4–5 indoors	trays/modules; 6 mm (¼ in) deep	21°C (70°F)	1½
Tomato, glasshouse	1–3 indoors	modules/trays; 18 mm (¾ in) deep	20°C (68°F)	1½

HERBS

Name	Month to sow/ preparation	Sowing method	Germination Temp.	Germination Time (weeks)
Burnet (*Sanguisorba minor*)	3–4	drills; 12 mm (½ in) deep	8°C (46°F)	2
Caraway (*Carum carvi*)	9	drills; 12–18 mm (½–¾ in) deep	10°C (50°F)	2–3
Chervil (*Anthriscus cerefolium*)	3–8 (successional)	drills; 6 mm (¼ in) deep	8°C (46°F)	2
Chives (*Allium schoeno-prasum*)	5–6	drills; 12 mm (½ in) deep	10°C (50°F)	2
Coriander (*Coriandrum sativum*)	4	drills; 12 mm (½ in) deep	8°C (46°F)	2
Dill (*Anethum graveolens*)	4–5	drills; 12 mm (½ in) deep	8°C (46°F)	2
Hyssop (*Hyssopus officinalis*)	4–5	seedbed drills; 12 mm (½ in) deep; transplant	8°C (46°F)	2
Lovage (*Levisticum officinale*)	4–5	seedbed drills; 12 mm (½ in) deep; transplant	8°C (46°F)	1–2
Marjoram (*Origanum* sp.)	2–3 indoors	pots/modules 3 mm (⅛ in)	10–15°C (50–60°F)	2

Name	Month to sow/ preparation	Sowing method	Germination Temp.	Germination Time (weeks)
Parsley (*Petroselinium crispum*)	3, 6	trays/drills; 12 mm (½ in) deep	5–24°C (40–75°F)	4
Summer savory (*Satureja hortensis*)	4	drills 6 mm (¼ in)	8°C (46°F)	2
Winter savory (*Satureja montana*)	4	drills 6 mm (¼ in)	8°C (46°F)	2
Sweet Cicely (*Myrrhis odorata*)	4–5	drills 6 mm (¼ in)	8°C (46°F)	2

The small black seeds of chives can be collected from the fading flowerheads in summer

Index

See also lists of garden plants from seed beginning on page 50
Page numbers in **bold** refer to illustrations